Raising Chickens:

An Introduction to Raising Chickens For Eggs & Meat

By

James Paris

Published By

Deanburn Publications

Contents Table

Table of Contents

Copyright

Introduction

Why Raise Chickens?

Economic reasons:

We live in a time when the world is facing 'Austerity measures' food shortages, and overall financial collapse of the banking system – so some believe.

While I believe that the banking total collapse may well be an exaggeration, the fact remains that we are all having to 'tighten our belts', in other words we are all looking for ways to reduce our daily bills by as much as possible. Keeping chickens for their eggs or meat can be very economical, and certainly very tasty!

Health reasons:

There has been a lot of publicity in recent years regarding the general freshness of the food on the supermarket shelves, with irradiation even used by some retailers to keep the food they sell fresher than longer.

This is however now in dispute, as the irradiation methods are claimed to do more damage than good to the consumer of such products. Not only this but the length of time that eggs for instance can stay on the shelves, leaves cause for concern.

If we buy for instance 'fresh farm eggs' just how fresh are they, and how long did they wait to get the 'sell by' date printed on? Do we trust out suppliers to do the right thing, or to make money – hmmm

Environmental reasons:
We have all seen the miserable conditions that battery reared chickens are kept in, most of us would like to see this changed if at all possible. Meanwhile if we have enough room to keep chickens we can do our own little bit so to speak.

There is also the recycling part of keeping chicken; although a lot of people do not realize this, chickens eat just about anything! They can certainly recycle household kitchen waste very effectively, and the kids love to feed them.

Children's education:
Rearing chickens at home and getting the kids involved in rearing them, is a great way to bring up your kids with an appreciation of the natural; world around them. I had chickens in our small-holding from about the age of 10 years old, and I still remember the lessons I learnt then about how to care for something other than myself!

In fact we were introduced to a range of animals including chickens, geese, ducks , turkeys and even pigs and cows. All these animals needed looking after, as well as providing for the kitchen table, they gave us a sense of well-being and achievement in our formative years.

Keeping Chickens

As I have mentioned there are many reasons for keeping chickens, as well as many things to consider when keeping chickens or any kind of livestock or poultry for food or monetary gain. Whether it be general health issues or the conditions in which the animals are kept, you must find out from your local authority just what the regulations are with regard to keeping livestock.

There are rules that apply country-wide and rules that apply state/authority-wide with regard to the well-being of your livestock, whether this be keeping chickens or any other animal . In general terms it is not 'rocket science' when it comes to animal welfare.

If you keep the animals well-fed and in a good clean environment as well as inspecting regularly for disease or the predations of vermin, then you will find that you will be on the right track. Do however check with the local authorities before embarking on any project involving animals.

Cruelty to animals is something that I personally cannot stomach, and indeed an individual who is cruel to the animals in their care need to punished within the full context of the law of that country. If an individual decides they would like to rear any animal then they must be made aware of the responsibility that this brings with it.

In the UK we used to have an advertising slogan "pets are not just for Christmas" because of the amount of animals that would be bought as Christmas presents, and then left to wander the streets when the owners could not be bothered looking after them.

That said, here is a few of the issues to address before keeping chickens in any enclosure that you may be considering.

A Coop for keeping Chickens in.

There is a lot of talk these days about free range eggs from free range chickens, and indeed it is always best – if not always practical – to allow the Chickens to roam free where possible.

However when it comes to roosting at night then the Chickens must have a place of shelter from the elements and protection from predators such as Foxes or Stoats. The Chicken hut or coop must have enough perches for your chickens to roost on so that they are able to keep off the floor.

When considering how many perches to put up consider around 10-12 inches (250-300mm) of space for each Chicken. This will vary of course with the type and size, but will give you a rough idea.

Keeping Chickens – Free range

What exactly constitutes 'free range' ? This will vary, but as a rough guide up to 800 chickens can be kept to one acre of ground and be classed as free range! However for most 'hobby' Chicken keepers I think that about 6-10 square yards per bird depending on the ground conditions, gives a good area for Chickens to prosper in.

Do remember that even 'free range' Chickens need shelter however so it is essential to have a Chicken Coop or Shed where they can roost at night, and where they

have nest boxes to lay in – and you can collect the eggs. If not, they will lay all over the place! In fact I remember as a lad hunting through-out the farm looking for the hidden laying places of the chickens.

A lesson in 'Perching'

The idea of 'Perching or Roosting' a Chicken rather than letting then stay together on the ground is not only to keep them of the cold and possibly damp floor, but also to stop them suffocating themselves! I remember an early lesson in this when I was a young lad and keeping chickens for the first time. If chickens have been reared under a lamp and do not have the influence of their elders so to speak, then they have to be taught to roost when they are very young – from about 6 weeks.

As an example of this..Every night I would go down to the Chicken coop and place about 100 Chickens on perches – this had to be done for about 1-2 weeks until they got the message that they had to do this themselves.

In the light of a small torch my girlfriend and I would place these Chickens on perches, it was a thankless job, especially when you had just placed the last one and turned to leave...when you would hear a 'plop' as the first of many jumped back off and it had to be done all over again, Arrggg ! There-after you had to go back at least twice to be sure that they were still sitting on their perch.

However one night I could only check on them once as we had a rare night out on the town planned, in the morning almost 50 of them were dead ! They had jumped back of the perches and had huddled up into a corner where the inside ones had simply suffocated – complete disaster, but a valuable lesson learned.

Some people would argue that chickens will roost naturally and do not need help; I can only say that **In my own experience** as just mentioned, this is not the case at all. This publication covers my own lessons learned in keeping chickens of many types, as well as drawing on the expertise of others. It is written by someone who has 'walked the walk' rather than by some academic who knows the theory but has never put it into practice.

Rhode Island Red – Special

Rhode Island Red – a special bird

The Rhode Island red (gallus domesticus) is probably best known by the public in general because of the striking pose that the rooster makes – not to mention the very loud wake-up call!

As the name suggests, the Rhode Island chicken is a deep red in color and is known as a utility bird which is raised

both for eggs and meat. Being a hardy bird the Rhode Island Red is popular with smallholders as it is easy to raise, and good in temperament.

History of the Rhode Island Red

It will come as no surprise that the Rhode Island Red was originally bred in Rhode Island and Massachusetts. It was bred from the Malay where it got its deep red coloring along with a strong constitution enabling the bird to survive and prosper in fairly harsh conditions.

The Malay cock from which the strain was bred, was actually imported from the United Kingdom and was a black breasted Malay cock which is now on show at the Smithsonian Institution as the father of the Rhode Island Red breed of Chickens.

Such an impact has this breed had on the public perception, that two monuments have been raised to pay homage to the Rhode Island Red; one in Adamsville in 1925 and another in Little Compton in 1954.
The Rhode Island Red Cockerel has been bred many times with the Sussex hen to produce prolific egg laying Chickens, thus enhancing their popularity amongst chicken breeders.

Rhode Island Red – Characteristic's
Generally the Rhode Island Red is a friendly bird, however the Cock can be very territorial especially towards strangers or youngsters. I remember my younger brother (who was a red-head) being chased regularly down the yard with this Rhode Island Red rooster leaping after him – we used to say it was his red hair that upset the Rooster!

It has been recorded that they have been known to attack and kill even fox's when agitated, however I do find this hard to believe (maybe you've had some experience here?). For these reason it is wise to watch their behavior with chickens of other breeds as they often do not mix well.

A sociable Chicken

All that said, the Rhode Island Red is a very good friend with those that it knows, and can become very attached to their owners. They will often respond to commands and will even walk alongside you like a dog or other domestic pet, if allowed to do so! For this reason also the Rhode Island Red is a popular bird – as everyone wants to feel loved and thought well off!

They are very much social animals and develop better when kept in groups as they look after each other and seem to enjoy company.

Rhode Island Red - Eggs

To summarize this article we look at the egg laying capacity of this great bird. The Rhode Island Hen is a very good layer and gives a large brown egg, often we used to get 'double yolkers' from the early birds – which was a great bonus to us kids, two eggs in the one shell seemed like a god-send!

A good well fed and healthy hen will lay 5-6 eggs per week, sometimes producing one a day every day which is far more than the normal average of around 4-5 eggs per week.
The Rhode Island Red – A 5 Star egg layer.

Chicken Notes

<u>Chicken Notes</u>

Other Chicken Varieties

There are indeed many varieties of chickens to choose from before deciding on the ones you want to invest in. Some chicken types are best for eggs for instance, some are better mothers than others, and some chickens are better reared for their meat, as they are heavy built.

It would be no use after all if you were to expend a lot of time and effort in making your chicken coops and buying feedstuffs; just to go and buy the wrong type of hen for your needs.

However the question of what are the best chickens for eggs, is not as easy to answer as it may first seem. The reason being that if a chicken is not content then it will not lay – end of story! So even a breed that is renowned for its good eggs – such as the Rhode Island Red – will not lay if the surroundings are poor or if the bird is stressed, under fed or otherwise unhappy.

That said, here are a few of the more popular chicken varieties chickens that may warrant consideration, if you are looking to populate your new (or old) chicken coop.

The Plymouth Rock

This is a very large bird that is good for the table, weighing in at around 9.5 lbs fully grown. It also is known as a dual purpose chicken as it is good for both meat and egg production. A very docile bird that comes in several varieties both white and barred, the Plymouth Rock Chicken also produces light to medium brown eggs and is regarded as one of the best chickens for egg laying.

The hens are known to make good mothers and will readily go broody. They are also known to be sociable and amiable, towards other pets around the yard.

Interestingly the Plymouth Rock is also a favorite of the fly-fishermen, who use the feathers for fly tying.

The White Leghorn Chicken

This breed is distinguished in that it is recognized as one of the highest egg yield chickens. Originally from Italy, it is also quite flighty (Italian blood ?!) and noisier than other chicken breeds. This chicken comes in a multitude of colors, around eighteen in all, and produces a clean white egg.

They are known to be exceptional egg-layers and can produces around 300 eggs per year if kept in ideal surroundings.

The Jersey Giant

This is a breed of chicken that is definitely more kept for its meat as it can reach an astounding 13lbs when fully grown. A popular bird around Christmas time as you may imagine! However a good sized egg for a healthy omelet!

The Java Chicken

Known as one of the 'old breeds', the Java has been used in breeding programs for decades. It is hardy and reliable and produces fine eggs. Weighing in at approx. 7.5 lbs. it also makes a fine bird for the table.

The Orpington Chicken

Another good bird for egg production, the Orpington is known as a big gentle bird that is quiet natured and easy to keep. Because of its docile nature it is known to be bullied by other hens so it is wise to keep a watch on it especially if newly introduced. Full grown at about 8 lbs, It

produces a large brown egg.

The Plymouth Chicken

This is a popular bird that has fine barred markings and is known to be easy to care for, producing a pinkish brown egg. It also makes a fine table bird weighing in at around 8 lbs.

Very similar in appearance and egg production to the Plymouth Rock chicken.

The Dorking

This is a lesser well-known bird but one that none the less is a great egg laying chicken that is also known as a good mother, and so is a good choice for a broody chicken if eggs are to be hatched.

Not a particularly large bird at around 6.5 lbs, it is known for the fact that it has 5 toes instead of the usual 4.

The Light Sussex

Very popular amongst chicken keepers, the light Sussex is a hardy bird and very docile amongst other species.

Producing a good sized tan to light brown egg, it makes a good chicken for eggs or indeed the cooking pot as the meat is renowned for its flavor.

This is of course not a comprehensive list of the chickens out there as there are indeed hundreds to choose from. Some will be familiar to you and others will be complete strangers.

Choosing the best egg laying chickens is probably a little academic really, as most of us will have no real say in the matter and will have a very limited choice according to where we actually live, and the supplies that are available.

Most chicken species are thankfully very hardy and provided they get the right amount of food, shelter and general care, will produce a good batch of eggs for 2-3 years.

Egg producing will of course decline as the chicken gets older and it is generally accepted that after around 3 years they are ready for 'the pot'; after this period they can be a little tough to eat.

Or if you are too attached to them they can still make great pets to have around and clean up the food scraps !

Keeping Chickens Secure

Securing a Chicken Run

If you are considering keeping Chickens, then one of the things you should know about is Vermin, and how to keep Rats and other vermin/predators away from your Chickens. The fact is that rats in particular will not just eat your chicken feed, but they will also attack and kill young chicks and even older hens that are perhaps out of condition.

Rats also spread disease and in no time at all can over-run an area if proper care is not taken to limit or eradicate them altogether.

The problem with Rats.

Rats are bad news for any poultry keeper for a number of reasons.
They eat your feed stuff, contaminating it with droppings in the process.

They chew and destroy anything they can get their teeth into.

Rats spread disease (think bubonic plague 13c!).

They will kill particularly any young chicks they have access to.

To add insult to injury Rats can carry two litters at any one time and so will soon over-run your chicken coop if they are not dealt with.

I think that you probably get the idea by now – Rats are

bad news, and have to be dealt with at the first signs of infestation.

Secure the area

Permanent chicken runs are different from mobile runs that you are able to move, in that you are able to make them much more secure by dropping the chicken mesh into a trench dug around 12-18 inches into the ground.

Make sure that the excess mesh at the bottom of the trench is turned toward you before filling it back in, this deters anything that would attempt to dig down and under the mesh. These precautions are necessary if you desire to build a chicken coop that is secure from predators such as rats, foxes , or even snakes depending on where you live.

The mesh must be of a type with holes no larger that half an inch (12.5 mm) and strong enough gauge depending on the predator that you want to keep out. I would rather spend a few extra dollars on a strong chicken mesh, than come home to find that a fox had broken in and killed all my hens ! It is false economy to build a cheap run at the end of the day.

I would also advise bringing the chicken mesh over the roof area of the run, if it is not already covered. This will prevent aerial attack from hawks or even crows (they will take away chicks), and will also prevent predators like the stoat or Weasel from climbing up and over the top.

Chicken coops - Preventing Rat Infestation.

Prevention is always better than the cure, so they say. To prevent infestation from rats is not 'rocket science'. At the end of the day, everything must eat to survive and so the simple way to prevent rats becoming a problem is to deny them access to food.

Ok, I'll acknowledge that this is sometimes very difficult, however your best attempt must be made if you are to be effective at all. Here are a few tips you can follow – just common sense really!

Keep all foodstuffs in sturdy (preferably metal) containers with secure lids that must be kept closed when not in use. Do not over-feed the chickens. This leads to food lying around on the ground that the rats will easily survive on. If you are scattering barley on the ground be sure that you are not overdoing it and that the chickens and not the rats will benefit.

Find and secure any access holes that the rats main gain entry by. I used to break a bottle inside any rat hole I found before filling it back in. Holes in walls etc should be filled with a suitable material such as cement mortar to be sure they do not dig through it again.

A constant program of rat-catching should be adhered to especially if you are in an area prone to them. Do be very careful with any rat poison and be sure to make it accessible only to your intended victims the rats.

I usually place it deep within a pipe or some such and make sure that there are no children for example, liable to gain access to it.
Rat traps, poison and chickens do not really mix well – use with caution !

Keeping Egg Production Going in Over Winter

There are a couple of good reasons why keeping chickens in winter or at least over the winter months, differs from keeping them over the summer or more moderate months of the year.

The main consideration for most people though, is just how much keeping chickens over the winter will affect the chickens egg producing capability.

Keeping chickens in winter – A short story!

My father first started keeping chickens for their eggs in the early 1950's, just after the war in fact. Food at this time was still rationed in the UK, and a government man used to come every day to collect the eggs from his small-holding.

When I say collect the eggs – that was the theory at least, however in the small croft that they stayed known as 'The Herricks' in Keith, Northern Scotland, the Glen suffered very bad winters.

Egg production was virtually non-existent, and so one day my father – who was still new to chicken rearing – decided that he'd have to do something about it.

Fact was that the Hens were standing around the yard on one leg simply freezing and obviously unhappy, when he enquired to some of the older hands in the Glen the answer was always the same "hens do not lay up here in the winter son ".

However, he was desperate enough not to take this 'good advice' and decided that he had nothing to lose by clearing out an old barn that he had on his croft, and putting the chickens inside for the winter.

Result? After only a couple of weeks his chickens started to lay 'Big Time' and he was the only man in the Glen producing any eggs in numbers; this was according to the government agent. Next winter they all had their chickens inside for the winter!

Keeping Chickens in Winter – here's how

Ok, to keep chickens in the winter and still produce eggs, is not rocket science! Here it is in simple steps.
Clear out or adapt an old barn if you have one, if not you will need to build a shed allowing for the number of hens you want to keep.

If you stick to around 5 sq yards per bird you will not go far wrong.

If possible keep one side open to the elements but able to close up if the weather comes from that direction. If the doors are closed be sure that you have adequate ventilation or airflow, this will help keep the shed fresh and free from disease.

Cover the floor with either wood shavings or straw. I have found that they are both very effective for the next step coming up. This will help keep the birds dry and off the cold floor.

Build a suitable amount of nest boxes for the birds, it's not an exact science but I've found that 1 nest box between 5 or so chickens is adequate for the task.
You want to keep the chickens busy because this in turn keeps them warm and there-fore productive. To do this I used to scatter a few handfuls of corn amongst the straw (only a few mind) as this would keep them scratching away for ages. Another trick is to hang up a few cabbages or turnips so that they are dangling within reach of the chickens.

Cold Chickens will not lay well!

This is to keep them pecking and busy, if there is nothing to peck at then they are liable to peck each other until the blood flows.

This can indeed lead to chicken cannibalism and can be very difficult to stop once it has started. Never allow this to go on, and if a chicken is getting bullied in this way then you must take it apart from the flock to recover.

For persistent offenders that peck constantly at other birds then they will have to be removed from the flock for some time to try and break the behavior pattern. If this does not work then the casserole pot beckons!

Keeping Chickens in winter or throughout the colder weather does indeed entail a little extra work. For instance the area under the perches particularly, has to be regularly cleaned out, as there will be a lot more poop to clean up.

The feed trays and water dishes also require regular cleaning out and topping up as part of the on-going routine.

Failure to apply this will result in a dirty environment which in turn will result in diseased and unhappy chickens which in turn leads to poor egg-laying chickens.

However I am confident that if you apply the preceding methods then you can indeed look forward to a good egg production over the winter months. Just remember the golden rule – if the chickens aren't happy….
They won't lay! If you are well organized and do not mind a little extra work then Keeping chickens over winter can be well worth your while.

Building A Chicken Coop

I consider building a Chicken Coop as one of the most rewarding outdoor projects that you can come across. Why? simple really; not only do you have the chance to build something that will ultimately be rewarding inasmuch as the eggs that will 'come forth' from it, but you will have an actual structure that your chicken's like and that you can gain pleasure from as in a job well done.

There is another aspect of building a Chicken coop and that is the chance to get others in the family involved. Kids love the fact that they are helping to build something to house some chickens – especially if you already have them ready to move in to their new home.

This is a real chance to build up some valuable bonding time with the family and produce a great food supplement with fresh eggs every day from your own Chicken Coop.

Believe me, the kids will love going out with you to collect the eggs every morning – it can be turned into a great games as well – how many eggs today do you think ?

A Mobile Coop

One of the first things that you must decide upon, is whether or not the coop has to be moved. Many coops are made to be mobile, as this enables the coop to be moved around an area to give better forage for the birds.

It also helps keep the chickens safe from predators, and out of the rose garden! This kind of coop is usually only big enough for around four chickens, as if it gets any bigger then it is difficult to lift it into position.

Also known as a chicken tractor, if it has wheels. Just where the name came from I have no idea !

This coop is fairly self- and is usually a modest affair made up from the simplest of timber shed designs. The run has an open floor so that the chickens can forage, and the covered shed has perches built in, and a nest box for the eggs that is accessible from the outside.

This design is very functional, and for that reason also very popular. Sometimes the floor itself can be covered with a wire mesh to prevent anything from lifting the edge of the run up and killing your chickens.

This did in fact happen to myself when I was a young boy, even though I had our Alsatian dog on the leash it was able to pull me forward and get into the run by lifting it up with its nose. The result was disastrous because we did not have a protective mesh on the floor – I was upset for days !

A Fixed Coop

A fixed coop is really any size of shed that you prefer, depending on the numbers kept. Unlike the formula for raising chickens over the winter above, and so leaving them enough room inside. The fixed coop does not have to be so large as they have got outside space as well.

Basically you have a shed with a doorway cut into the end near the bottom, this leads into the wire mesh area where the chickens get out to feed. This coop design has all the same features as the mobile coop, with regard to nest boxes and perches.

The main difference is that it is in a fixed position and so you can erect it with all the above cautionary measures mentioned with regard to rats and other predators.

Plans for many kinds of chicken coops can be found in the resources page at the end of this article.

Feeding

There are many automatic feeders available today, for both grain and water, so I will not go into much detail here. Perhaps I will put a link on the resources page for this equipment also.

Suffice to say that you do not have to feed the chickens every day if you do not want to, as the automatic feeders available allow for many days feeding before needing filled up again.

Even the water dishes can hold many days water and are designed not to go foul as a result. That said, I would always check that the water in the bowl is clean and fresh. The last thing you want is sickness brought on by contaminated water.

A top-tip to keep the hens from picking one another is to hang up cabbages or other veg from a rope set to chicken-head height.
This will keep them pecking on the veggies rather than each other.

Keeping Healthy Chickens

In general it has to be said that chickens are extremely hardy creatures, and in all the time I have kept them I have seldom had any health issues with them.
The main thing is to keep them in as clean an environment as is practicable – dirt will not harm them; filth will. It is as simple as that.

Keep them on a good nutritional diet and allow them a place where they can scrape about, and chickens are as happy as a pig in muck!

As I have mentioned earlier, chickens will not lay consistently if they are unhappy in the least, and one of the first signs of sickness is that they will stop laying.

Rather than a self-diagnosis which will probably be wrong, I would always advise calling in a vet in the event that your chickens do get sick for any reason. Especially with the scares about avian flu doing the rounds, it is better to be safe than sorry.

With that said however, here is a short list of the most common ailments to affect domestic chickens.

Northern Fowl Mite: This is a mite that lives and feeds on the chicken, staying close to the skin where it draws blood around the clock. Reddish-brown in color, it leaves the feathers of the bird looking discolored due to the waste of the mite and the birds overall condition.

If left untreated, this will lead to lethargy, loss of egg production, and in severe cases death.

Chicken Mite: This is a small mite yellowish in color, that feeds on the chicken usually after the hours of darkness. It can live in the nest boxes and the perches, and come out to feed on the chicken at night time. The coop as well as the chicken has to be treated to remove this threat.

Scaly leg Mite: This is a mite that appears on the feet and legs of the bird, separating the scales on the legs and living in the cracks in the flesh. This leaves the feet and the legs of the bird very tender and sore.

Poultry Lice: These come in many shapes and sizes, but work by chewing with their mouths rather than sucking the blood as is the case of mite infestation. They feed on the dry skin of the bird and cause irritation rather than blood loss. The results however are similar; loss of appetite, egg production and overall health of the bird.

Treatment:
Fortunately provided these events are discovered early on, treatment is fairly straight forward most of the time. There are in fact many products on the market today, and this is where a visit to your vet or even local animal produce store will help you.

However a popular treatment for mites is Sevin dust. This can be used on the bird and the coop to kill the mites. Eprinex is a liquid treatment that is also very effective.

With scaly leg; petroleum jelly, linseed oil or vegetable oil rubbed into the leg every few days until the skin is 'smooth' again, will usually sort out the trouble.

Prevention:
As usual, prevention is always easier and more cost effective than the cure. The chances of you getting one of the above or indeed other chicken ailments are fairly high. This however is not a great drama, and the main thing is to keep the environment around your chickens clean, especially the coop itself.

This should be regularly washed out with a good strong cleaning fluid, to prevent infestation of the perches and nest boxes. Animals sometime get sick – it is a fact of life. By taking simple measures however, you can prevent things from getting out of control – and keep egg production high with good healthy chickens.

●●●

<u>Chicken Notes Page</u>

<u>Chicken Notes Page</u>